BLUEPRINT FOR WORLD PEACE

PAUL BREER Ph.D.

Quantum Discovery
A LITERARY AGENCY

Chat GPT contributed to the compiling of data and
made suggestions for China–U.S. cooperation.

The front cover painting (Metamorphosis) is by Varouj Hairabedian.

ISBN
979-8-89641-103-1 (Paperback)
979-8-89641-104-8 (eBook)

Table of Contents

Introduction

In the summer of 2025, we are once again stumbling toward the precipice of war, this time a full-blown nuclear conflict which threatens to leave millions dead and the planet in ruins. In Moscow, Vladamir Putin trumpets his intention to restore the 18th century empire of Peter the Great, while in Beijing, X Jinping grooms his forces to take Taiwan by force. In the Middle East, Iran and Israel hurl missiles at each other as the United States stands by with aircraft carriers and bunker bombs at the ready. Meanwhile India and Pakistan, both nuclear nations, trade tit-for-tat incursions into Kashmir while in North Korea, Kim Jong Un starves his people as he stockpiles weapons to keep enemies at bay. And then, most unnerving of all is the rising tension between China and the U.S. in the South China seas where daily acts of intimidation have brought us to the brink of outright conflict.

Keep in mind that we humans have been in situations like this in the past—in fact, so many times that a list of wars significant enough to deserve names takes a whole page.[i] Explanations of what precipitated those wars are superficial in that they fail to look deeper than what can be caught by the casual eye. While each conflict has its idiosyncratic features, a closer look suggests that there may be a more basic issue at play. And it is our failure to address that issue that destines us to have more wars, each more devastating than the last, until we manage to destroy civilization altogether.

Recall that as a species homo sapiens has been around for at least 300,000 years, during which time we have failed to achieve a sustainable world peace. Despite the unbelievable progress we have made in science (astronomy, biology, and physics), technology (cars, airplanes, computers and television) and the arts (literature, music, painting, and architecture), the way we interact with each other remains primitive. While that is troubling in the way we treat citizens of our own country, it rises to a dangerous level in our attitudes toward countries with cultures markedly different from our own. It is no hyperbole to say that it is our paranoid attitudes toward "others" that have once again led us to the brink of world war.

The 20th century, the century of WWI and WWII, Hitler and the Jews, Stalin and the kulaks, the genocides in Armenia, Rwanda and Bosnia, the Cambodian killing fields, Pearl Harbor, and Hiroshima, stands as one of the cruelest on record. Despite recent advances in international cooperation (e.g., the United Nations), it is clear that the way we perceive and interact with those in other parts of the world has improved little since the Stone Age.

And the problems continue even to the present. Just look at what happened a few years ago in Myanmar and South Sudan. The suffering brought about by ethnic cleansing defies belief. Tens of thousands of innocent civilians were forced out of their homes, raped and murdered—then to make sure the ones who escape didn't return, their villages were torched to the ground, all in the name of ethnic purity. Years earlier the Hutus of Rwanda perfected the cleansing model used—keep denigrating the other (e.g., by flooding the airways with the message that they are "cockroaches") until they are no longer seen as human. Then they can be butchered with impunity.

As the carnage got acted out, the United Nations sat by, lamenting the horror, but unwilling to stop it. It was prevented from acting by a clause in its charter which prohibits it from "interfering" in the internal affairs of a sovereign nation.

Some would say that refusing to interfere when thousands of innocent people were being driven from their homes is difficult to defend ethically. We allowed it to happen—and then wrung our hands as we watched the atrocities unfold nightly on television.

While most of us in the U.S. accept our own federal government's right to intervene in state matters—e.g., to guarantee minority voting rights, we refuse to give a similar right to the U.N. to intervene when there is a humanitarian crisis in one of its member nations. And so, we are rendered helpless. In ethical terms, the United Nation's refusal to stop the Myanmar government from killing and expelling its Rohingya minority would be equivalent to Washington's refusing to interfere if the state of New York began killing and raping its residents of Italian descent, then forcing those they missed to flee into neighboring Rhode Island. We wouldn't allow something like that to take place in our own country—yet, we turn a blind eye when the equivalent happens internationally. In each case the guiding principle is clear; no matter how appalling the situation is, sovereignty comes first; compassion, if heard at all, plays second fiddle. And that is a recipe for eventual disaster. It has happened before and unless we reverse gears, it is going to happen again.

We are like a herd of ice age mastodons stampeding toward the cliff's edge, driven not by the spears and arrows of men hungry for our meat and hides, but by our hostility toward those whom we perceive to be different. The mastodons had no choice as they lurched toward oblivion, sides ripped open and trunks aflame with the smell of death until they fell to their death on the rocks below. Unlike the mastodons, we humans have a choice. It is true that we are being propelled, but it is not by an external force over which we have no control. What is driving us closer to our doom is the hostility within us, a hostility toward people whom we see as "other" than ourselves.

To return to our previous question, why is there a disconnect between our advances in science, technology and the arts on the one hand and the way we treat our neighbors on the other? While we have done so well in other aspects of life, why have we made so little progress in our relationships with those from cultures different from our own. In a word, what is it that over the centuries has driven us from war to war? The typical explanations do not satisfy. With recent wars in mind, some will point to the assassination of a diplomat (WWI) or the rise of a charismatic leader (Hitler) or to the demands of an oil-hungry Japan (WWII). But such answers simply raise further questions. Why did we, the heirs of Bach and Mozart, Shakespeare, Rembrandt, Dante, Michelangelo, Tolstoy and Dostoevsky, comply in the slaughter of millions that followed?

By now it should be clear that the root cause of that compliance lies deeper than any specific event. Its nearly universal character points to something hidden from consciousness—a distorted image of self and others that lies deep in the human psyche where it is coiled to strike whenever we feel threatened. Because of its volatile nature, that distorted perception remains beyond the reach of reason—making it impossible to live peacefully with each other. Until we take steps to change the way we define self and "other", like the mastodons of the ice age, we are destined to be driven over the cliff.

1

Identity and Politics

The solution to our problem lies in changing the way we define who and what we are—in a word, our identity. Sometimes we use the word to define the way we typically think, feel and act—in other words, our personality (he's shy at parties, she's quick to anger, he tries to be funny but really isn't, she likes to mother people, etc). At other times the word identity refers to the social categories we belong to (he's American, she's Danish, he's upper middle class, she's blue collar, he's black, she's white, etc.). While both psychological and social identities are important in life, this essay assumes that it is the social categories that represent the greatest threat to world peace.

The way we create and cling to our social identities has a direct impact on the way we feel about others, and, in turn, the way we relate to them. If I define who I am as a separate, unique inner "I", everyone else (through extension) becomes an "other" to me. The more "other" they are, the less I trust them. ("I am Hutu, you Tutsi—you are different from me; I don't trust you.") For most of us life is a struggle in which our most important goal is to preserve our identity as an individual self. And that constitutes an open invitation to suffer. In The Book of Not Knowing, Ralston spells out the connection: "...recall any form of suffering, any distress, worry, upset, fear, misery, stress, longing...and consider long and hard: if you didn't care about *you* persisting in any way, if it didn't matter to you if you existed or not…, would you suffer any of these things? The answer is no. You cannot suffer when there is no self trying to survive."[ii]

While self-preservation may not be the only source of our suffering, it is by far the most common. But there is more to the issue than meets the eye. The self we've been referring to consists of more than a core sense of "I-ness", i.e., the experience of being a unique, solitary individual. As we grow up, we add all kinds of things to that core. As Tolle reminds us in The Power of Now, early in life a child begins to add possessions to his basic identity. His favorite toys, for example, become part of who he thinks he is. They are *his* toys; any threat to that beloved teddy bear he clings to every night is interpreted as a threat to himself and will elicit an angry response equal in intensity to being assaulted physically. Along with books, movies and other toys, the teddy bear has become part of the way he defines who he is.[iii]

In time he will add social categories like family, gender, race, occupation, social class, religion, and nationality to his identity, much the way an onion core adds layer after layer as it grows. With each new layer of identification, he takes on the responsibility of promoting and defending the whole onion. By the time he reaches adulthood, even his children can become extensions of his self. Sometimes this is pleasurable, for example, when one of his kids makes the honor roll. It's not so good when the opposite happens, like when a neighbor brags about his child's achievements at school and he can't say anything good about his own child in response. Now that he's a parent, his children have become a part of the way he defines himself; praise for them feels like praise for himself; similarly, a threat to them is experienced as a personal threat. The children have become part of his self-identity.

Family represents one of the first layers to be added to the identity onion. Chances are that there will be many more to follow, each offering new opportunities for both happiness and despair. As we mature, it seems perfectly natural to identify with our gender and race; identification with occupation, social class, nationality and religion come later, depending on how important they are in our particular culture. From a political point of view, the most critical layer we add to our identity is our nationality. If I were to ask you to answer the question, "Who am I?", I'm sure that somewhere in your list of identifiers would be "I am an American." Like religion, it is an identity we adopt in childhood, clarify in adulthood and cling to until the day we die. Once we expand our sense of self to include the nation we live in, any threat to our country will be felt as a threat to ourselves and will give rise to self-righteous indignation. If, for example, a competing country imposes tariffs on our exports or if one of our diplomats is accused of spying or, heaven forbid, if a plane from another country should accidentally invade our airspace, we would automatically howl in protest and seek ways of retaliating.

Identifying with a particular religion can precipitate an equally angry response. Witness how Muslims react to caricatures of the Prophet or how fundamentalist Christians feel when football fans mimic Tim Tebow in one of his Rodin-like poses. What makes these threats so explosive is the way we interpret them. We see them as attacks on the way we define ourselves, i.e., as threats to us personally. As such, they arouse a response in us similar in intensity to the wrath we feel when someone attacks a family member or when our gender or race is disrespected.

True believers (particularly religious or political ones) represent our gravest threat to world peace. Witness the ravings of a Hitler who saw the Aryan "race" as rightful rulers of

7

the world or the teachings of Shariah law which commands Muslims to kill anyone who converts to another faith. Opinions can also serve as add-ons to the self-onion; they needn't be religious or political in nature. Take for example, the belief that Peyton Manning is the best quarterback in the history of the National Football League—better than Tom Brady, Joe Montana, Sammy Baugh, Y.A. Tittle etc. Aided by a few drinks, grown men have been known to come to blows over that simple assertion. On the surface it looks like nickel and dime stuff, but underneath our very lives may be at stake.

As long as we see ourselves as fragmented egos whose overarching goal is to promote and protect our personal identity *and all its extensions*, there is little hope of achieving peace at either the domestic or global level. We may continue to marvel at the latest discoveries of astronomy, quantum physics or molecular biology while enjoying our favorite plays, concertos and novels—but we will never be free of tension or find lasting peace internationally until we change the way we define who we are.

The trouble, of course, starts with mistaking the concept of an inner "I" for a real entity—and then absorbing into that illusory entity things like our family, race, gender, social class, occupation, and so on. When one of the things we identify with is attacked, we interpret it as a threat to the core of the onion itself, that is, to ourselves. This is where the explosiveness originates—family, gender, race, class, religion and country being mere extensions of that core. Extensions or not, some of these layers have the capacity to incite riots or murder—and in the case of religion, political ideology and nationality, to start wars. (*pause*) Krishnamurti was apparently thinking along a similar line when he wrote the following:

> We recognize that the self is in constant flux
> yet we cling to something which we call the

permanent in the self, an enduring self which we fabricate out of the impermanent self. If we deeply experienced and understood that the self is ever impermanent, then there would be no identification with any particular form of craving, with any particular country, nation or with any organized system of thought or religion, for with identification comes the horror of war, the ruthlessness of so-called civilization.[iv]

Krishnamjrti is saying that the heart of the problem lies in our insistence on perceiving the self as a permanent entity rather than as a changing pattern of personal traits. And it is that mistake in defining who and what we are that draws extensions like country and religion into our self-definition— and ultimately, into the "ruthlessness of so-called civilization." We may agree that Krishnamurti is right in that we do change slowly as we age or experience major jolts to life, but we insist that the underlying person we identify with remains the same, in other words, that we remain the same person we have always been even as that person *undergoes* certain changes over time. And that way of thinking is the problem. There is no permanent agent tucked away inside the body which *undergoes* change but never changes itself—and thus, no permanent onion core to absorb extensions like religion and nationality to its self-definition as the body-mind ages. The changes may be real, but the self thought to be undergoing them is an illusion.

Regarding wars, the only way we can eliminate them and all the suffering they cause is to stop making certain extensions part of our personal identity. With respect to religion, we have to become seekers after understanding, rather than defenders of the one and only true faith. Regarding nationality, we have to become citizens of the world rather than patriots of the U.S., France, Egypt, China or Somalia. Neither of these is likely to happen so long as we continue to absorb any kind of ideology into our self-identity.

You may argue that even if people didn't identify with their country, they would still fight over resources like oil, gas, grazing land or fishing rights? Perhaps—but even in fights over resources, identification of a sort plays a role, e.g., the land is ours, the oil belongs to us, our people have lived here for generations, etc. It would be an overstatement, however, to argue that identification with one's religion or country (where that represents an extension of the self-onion) constitutes a sufficient explanation for why war is still with us. It's certainly possible to think of oneself as a Muslim, Christian, Egyptian or German without taking up arms against those who share a different faith or nationality. A complete explanation for war would have to include a provocation by one religion or nation toward another—but, of course, history tells us that it doesn't take much to get things started (e.g., witness the trivial origins of WWI).

By their very nature, religion and nationality, once absorbed into one's self-definition, make one hyper-sensitive to the actions of anyone whose beliefs and loyalties lie elsewhere. To the extent, for example, that Christians and Muslims seek to convert the other to what they believe to be the only true faith, the atmosphere between them will be saturated with hydrocarbons, ready to burst into flames at the drop of an offending remark or deed. All it takes then is a readiness to respond aggressively to frustration—a readiness that appears to be lodged in the DNA of most human beings. (*pause*). A similar argument can be made for the tension created when we define ourselves (at least in part) as members of a given country; while in and of itself that identity needn't lead to hostility toward members of other nations, by virtue of its personal nature (e.g., I am a North Korean, you are an American), it provides enough combustible material to inflame a whole continent if and when conflict arises.

So, what can we do about it? The long-run answer lies in giving up all forms of identification—that is, becoming what Ralston describes as "nothing in particular." But that requires gaining access to our fundamental nature—Pure Consciousness, something that may not happen on a broad scale for decades or even centuries. Short of that ideal, we can make progress by substituting a broader identity for a nationalistic one—e.g., substituting "I am a human being" for "I am an American" or in the case of religion, substituting "I believe in a reality beyond this material one." For "I am a Muslim." That could change the way that people living in the U.S. feel toward citizens of China, Russia or North Korea—or those who cling to Islam, Christianity or Buddhism—all of whom can now be seen as members of a common family, the family of *homo sapiens*. Of course, members of the same family often quarrel, and that will be true in our relationships with citizens of other countries and beliefs, but without the stigma that comes with perceiving foreigners as "other than me", those problems should not be difficult to handle.

Perhaps the importance of self-identity in international relations can be made clearer by imagining what it would be like if we were forced to confront a powerful force from outer space. In some ways, we'd feel much the way our colonial forefathers felt when they were forced to confront the British army. Up to that time, each of the thirteen colonies fended pretty much on its own. Each had its own militia. Even when George Washington convinced them they could only win if they agreed to work together, they insisted on naming their own generals (each colony was limited to two). After four years saddled with one defeat after the other, the continental army under Washington won a victory in New York, bringing the war to an unlikely end. A few months later the army turned over authority to a convocation which proceeded to write a constitution and declaration of independence.

While this American experiment in federalism may give us insights into what could work at the global level, it does not constitute an adequate model for an improved version of the United Nations. What's missing in the 21st century is something to take the place of the British army, i.e., an oppressive force that can be defeated only if separate entities are willing to give up some of their sovereignty and agree to work together.

The force most likely to play that role today would be an invader from somewhere in our galaxy. When faced with such a force, we'd quickly put aside the problems we were having with our neighbors and concentrate our energy on the more dangerous threat at hand. Our new sense of solidarity with our fellow homo sapiens would allow us to forget about our territorial conflicts as we prepared to battle the aliens from outer space. But here in the 21st century such a threat is unlikely to appear anytime soon, which means that if we are to surrender some of our sovereignty to a central authority, we need something else to make us do it.

As I suggested above, I think the answer lies in something as basic as the way we define ourselves. If we define who we are as separate, unique "I"'s, everyone else (through extension) becomes an "other"; the more "other" they are, the less we will trust them. And that is a recipe for territorial paranoia—and an endless series of conflicts..As long as we see ourselves as autonomous egos whose overarching goal is to promote and protect our personal identity *and all its extensions*, there is little hope of achieving peace at the global level.

But some of you will protest. Don't we still have to be somebody, you say—a separate person, someone different from everybody else, someone with a distinct character? And if that's true, how can we prevent people from saying "I'm an American" or "I'm a Russian"? But that assumes we will always be saddled with the mistake of defining ourselves as an inner "I" with contra-causal powers. But what if that is not the case? How would individuals who have successfully dispelled that illusion feel about their country or religion? Would they be as zealous in their patriotism or as fanatical in their faith as those who have absorbed nation and religion into their self-image? If they no longer identify with an unchanging inner "I" but take Krishnamurti's advice and perceive themselves instead as an endlessly changing pattern of traits, would they continue to *define* themselves as Christian or Muslim? (*pause*) I think not. If they've managed to give up their core identity, what's left for them to identify with? Anything at all? If there's no permanent inner "I" left, it doesn't make sense to get all worked up over what social category that "I" belongs to. The body-mind remains, but there's no longer a permanent, internal entity thought to possess the body, thus, no ego to be concerned about.

So, let me rephrase my question: in the absence of ego-concern, what would our lives be like? (*pause*) Wouldn't they be filled with one sensation after another—a sound here, an image there, the feel of something in your hands followed by the smell of something in the air—a sequence of present-moment experiences interspersed with an occasional problem-solving thought? In that kind of sequence, there is no "you" left—a body-mind, yes, but no inner "I" or "me" that is concerned about *what* it is—or what others might think it is.

And because there is no inner "I" left to worry about, there is no need to flesh out that identity with attributes that tell the world *who* that "I" is (what gender, race, religion, nationality, etc. the "I" belongs to)—and thus no need to promote or protect the groups in which the "I" holds membership. It is only when we identify with an inner "I" and absorb into that definition our belonging to such groups that we become emotional in their defense. Give up the core illusion and its extensions like nationality and religion quickly wither away. For those who have succeeded in dispelling the illusion of a permanent, separate self, sovereignty loses its sacred character; by extension, the same can be said for religious crusades, ethnic quarrels, armament races, trade wars and patriotism in general.

Piecemeal measures that foster cooperation among nations (e.g., the European Union, NATO, NAFTA, and the U.S.-Russian nuclear arms treaty) will help, but fall short of what is needed. They fall short because they leave the basic cause of tribal hostility untouched. *The root cause of that hostility is the error we make when we define ourselves as autonomous, free-willing egos — and absorb into that identity the social units (particularly nation and religion) to which we belong and which we feel obliged to defend.* The only long-term, permanent solution to that hostility and the wars it engenders is to change the way we define who we are. Only when we have thrown off the yoke of self-identification will we be ready to forego patriotic and religious fervor—and surrender sovereignty to an international authority that has the exclusive power to act militarily. I don't see any other road to peace. We may never get to where we need to go, but our best hope of finding peace is to keep such an ideal in mind as we stumble ahead.

While science will certainly make it harder to believe that each of us possesses a divinely-given soul, it remains unclear whether the new skepticism will erode our faith in the individual altogether. And if it does, what's going to take its place? It's likely to be something new, something secular and uniquely Western—perhaps a society with less concern for personal autonomy and privacy and with greater value placed on group membership and collective goals. As our belief in free-will crumbles in the face of advances in artificial intelligence and genetic engineering, public concern will begin to shift from efforts to protect the sanctity of individual choice to programs for altering the determinants of behavior. While we already acknowledge that genes and environment play a role in shaping human experience, in the future these two factors will assume center stage in our political and economic deliberations.

Once we accept that all behavior is determined by previous conditions rather than by an inner "I" that can make uncaused choices, our attention will turn to discovering ways to remedy the conditions that lie at the roots of unwanted behavior—e.g., the biological and socio-economic conditions that spawn disease, aggression, ignorance, intolerance, alienation, crime, and premature death. Concern over individual rights will continue to play a role, but without a belief in the sanctity of individual souls, those rights will no longer be conceived as *inalienable*, that is, as claims given by either God or natural law—i.e., rights that cannot be taken away. If those rights are to be justified, as I hope they will, it will no longer be on religious grounds, but on the secular principle that they contribute to human happiness.

In a postmodern world where we no longer identify with a master controller somewhere between our ears, our primary concern will shift to those background factors acknowledged to be the determinants of the thoughts, feelings and actions that arise in our body-mind. A change of that magnitude would amount to a major upheaval not only in the way we see ourselves, but in the way we relate to each other. While the details of that revision remain to be worked out, they are likely to be based on a more realistic assessment of what it means to be human. And that promises to be a good thing.

2

Dissolving Territorial Paranoia

When I talk about territorial paranoia, I am not talking about the way we Americans view citizens of the United Kingdom or any of the other English-speaking countries we have contact with. Our attitudes toward "other" peoples varies with how "other" we perceive them to be—and in the case of U.K., Canada, Australia or New Zealand, that sense of "otherness" is minimal.

But for a more relevant example—one that speaks to the present moment—consider a relationship that is taut with negative energy, one that could easily spill over into a disaster of unimaginable proportions. I am referring to the relationship between China and the United States. These are the two richest and most militarily formidable nations on earth, and it is precisely this very fact that makes their mutual "otherness" one of the gravest threats to world peace.

Yet, this very tension conceals a profound and hopeful possibility. In this pivotal chapter of history, the United States and China stand not just as rivals, but as nations uniquely positioned to redefine the global order. That redefined future will be shaped not by rivalry or isolation, but by the ability of these two nations to rise together and assume their role as architects of a new era. As the rest of the world watches, their ability to transcend cultural differences will potentially spread to areas of the globe where conflict between nations threatens to explode into war. That, at any rate, is the theory.

With that in mind, it makes sense to investigate the two cultures more carefully—with an eye to understanding what it is that makes them so different—and using that understanding to relieve some of the tension between the two. Whatever we learn from that investigation can be used to improve relationships between countries more broadly.

Let's start with some basics. There is a verbal dichotomy that captures what is probably the most fundamental difference between the two cultures. I refer to the distinction between collectivism and individualism. For centuries China has considered the well-being of society more important than the rights of its individual members. Harmony in government, above all, is considered sacrosanct. Compare this to American politics which over the years has often been fraught with yelling, screaming and the exchange of personal attacks. The Chinese ideal, based as it is on a combination of Confucian, Buddhist and Taoist beliefs, prizes imperturbability and the building of mutual trust. By contrast, American politics has been shaped by a revolt against medieval religious beliefs. That revolt lingers on today in the form of arguments for and against the separation of church and state.

In both China and the United States, the political structure of what we see today owes its provenance to a set of religious and secular beliefs that were adopted in the past. In China, the accent on harmony that originated in religious ideology continues to place its stamp on the political structure we see today, namely one where collective values are given priority over individual rights. In America, the social system that emerged from the Enlightenment of the 18th century persists in the priority given to personal freedom, self-expression and individual rights, all enshrined in the U.S. constitution adopted in 1776.

While collectivism and individualism remain hallmarks of the China-U.S. relationship, there are growing similarities between the two cultures as well, in particular the decline of formal religion. In China what started with the replacement of religion with Marxism under Mao continued with the secularization of Chinese culture under Deng. Today, the Chinese are less attached to traditional ancestor worship or Confucian rules of etiquette than they were decades ago. This is especially true for the young who exhibit a growing interest in secular ideas broadly and Western materialism specifically. In the U.S. youths are increasingly "spiritual but not religious" as they move away from organized religious to explore Eastern practices such as yoga and meditation.

There are several other cultural issues that need mentioning. Despite the growing income inequality in America (not to mention the increasing number of billionaires in China), when it comes to legal rights and everyday interactions, Americans are more egalitarian than the Chinese. The contrast shows up in the greater respect the Chinese bestow on authority figures and their elders, compared to America where authority is begrudgingly accepted and paid professionals handle most of the responsibility for elder health care. Important too is the growing concern about the environment where concern can be seen in both China and the United States, particularly among the young. Climate warming is increasingly viewed as a global problem than can be fixed only if the countries contributing most of the carbon dioxide (read China and the United States) work together.

Globalization clearly has the effect of blurring cultural boundaries. In China young professionals are adopting Western lifestyles, self-expression, and career-first mentality while in America youth are turning to yoga, mindfulness, acupuncture and an interest in harmony with nature. Chinese spirituality remains but it is integrated into daily life where it is more about "living well" than achieving salvation. In America religious interest has more to do with faith, the Bible, moral choices and sectarian doctrine.

These developments among Chinese and American youth are encouraging and should be paid attention to even as the broader divisions of collectivism and individualism remain basically unchanged. As an AI source puts it, "While the Chinese model still leans toward *collective stability and pragmatism,* deeply shaped by its Confucian-Taoist-Buddhist roots, the American model still leans toward *individual freedom and rights,* deeply shaped by its Christian-Enlightenment heritage."

In summary, while the basic differences between China and the United States remain strong, there are changes currently taking place in both cultures, especially among the young. If we want to ease the tension that now exists, a tension that constitutes a threat to global peace, it is with the young that we should concentrate our efforts.

3

Getting Specific

If our goal is to achieve and sustain a state of peace in the world, what concrete steps can we take to move in the right direction? At the present, we are light years away from reaching a goal of peace. A seemingly endless history of war, death and destruction makes it clear that if homo sapiens is to find enduring peace, it can only be done if nations agree to disband their armed forces and accept a bold, new structure in which the United Nations is given a monopoly on military power. We already have a version of that organization, but it is so hampered by constraints that we are helpless to stop countries from resolving their problems via political violence.

And so, once again we find ourselves teetering on the brink of world war. Strengthening the United Nations (or replacing it with a more effective body) will entail major changes in both the structure of the organization (representation, voting rights, financing, etc.) and the rules regarding what it can and cannot do (intervening in state affairs, migration, secession, etc.). Implementing such changes will probably take the better part of 25 years. If that seems like too long a time to spend in reaching the goal of global peace, the magnitude of the task requires no less of an effort. The alternative is to continue unchanged along the path we are now on—and get ready to accept the consequences, among which another world war is most likely. In the meantime, let's take a look at what can be done right now.

Military spending

According to the Stockholm International Peace Research Institute (SIPRI), total global military spending in 2023 was approximately $2.44 trillion USD, the highest level ever recorded. That year, the top spenders were as follows:

United States–$916 billion
China— around $296 billion (estimated)
Russia— around $109 billion
India— around $84 billion
Saudi Arabia— around $75 billion

Other countries like the UK, Germany, Japan, and South Korea also spent heavily.

This is an astonishing amount of money. Think of what it could be used for if it were not considered necessary for countering attacks from foreign governments—or lest we forget, for launching an attack of one's own. It will be argued, of course, that all that investment helps to raise an economy's GDP—regardless of its intent. True, but directing that money to projects that reduce suffering in the world while at the same time raising GDP via payments for labor and equipment constitutes a more civilized alternative.

With that in mind, a first step would be for the United Nations to persuade member states to agree to a 50% reduction in military spending over the next five years—and then to repeat that for the next three five-year periods at which time military spending in each country will be required to fall to zero. Inspectors from the U.N. will verify compliance with the order; strict penalties for non-compliance will be applied.

Such a dramatic shift in policy will be most difficult for countries like the U.S., China, India and Russia that spend billions each year on military personnel and equipment. But there are obvious benefits. Once accepted, all military expenses would be eliminated completely, saving the U.S. and other countries trillions of dollars annually which could be spent on paying down national debt, reducing poverty, building houses for the homeless or improving education.

Assuming that the 50% reduction is repeated until all countries have reduced their spending on national militaries to zero, a phased-in transfer of equipment to the U.N. can begin. Once all planes, bombs, missiles, tanks, ships and armored vehicles have been turned over, each nation will be expected to send personnel to a U.N.-sponsored global armed service and pay taxes for their upkeep, while maintaining its own domestic police force and national guard.

The goal of the arrangement is to take the capacity to make war out of the hands of individual nations by assigning exclusive military power to an international agency. The mere suggestion of such a policy is bound to elicit an hysterical response, but if we are serious about wanting to live in a peaceful world, this is what we have to do. While such a radical proposal remains impossible in today's world, it will sound more reasonable once we stop identifying ourselves as patriots of a particular society or followers of the one true religion.

Student exchange program

According to an agreed-upon plan, each year China will send 1,000 of its recent high school graduates to spend a year with a family of similar socio-economic status in the United States while the U.S. sends an equal number of its youth to study and learn in China. The year abroad will involve the learning of each other's language and customs as well as trips to historic sites and national parks.

There are several expected outcomes, each of which will serve to strengthen ties between the two host countries. Most importantly, the program will improve participants' understanding of the host culture and make them more likely to protest when they return home and hear derogatory things about that country. Among adults it will lead to an increase in Chinese visits to the U.S. and reciprocal visits of U.S. citizens to China, a change which will not only benefit each country financially, but allow visitors to ground their opinions about the other country in actual experience. There's another benefit involved that may not be so obvious—namely that with all these young people living abroad for a year, the governments will be less inclined to fire rockets at each other; after all, their own kids might be living there.

Participation in the exchange program will be strictly optional and fully-funded by the two governments. The program will be repeated each year until the nations agree to stop.

Cultural exchanges

As interest in each other's culture improves, China and the U.S. will resume exchanging concerts, plays, and art exhibits.

Global Warming

i. The two governments will create a Joint China-U.S. Climate Science Task Force to pool climate modeling and satellite data. That data will be used to build early warning systems for drought, flooding, and rising sea levels.

ii. China and the U.S. will create a Green Energy Co-Research Program focusing on nuclear fusion, battery storage, and carbon capture technologies.

iii. China and the U.S. will co-host an annual Climate Summit focused on the financing of global decarbonization initiatives.

Space Exploration

i. Initiate a multinational space program for a joint Mars mission with the U.S. and China as anchors.

ii. Write a Space Environmental Treaty in which China and the U.S. agree to (a) outlaw the weaponization of space and (b) manage the distribution of space debris.

Elimination of Poverty

i. China and the United States will create a bilateral development fund to support programs to alleviate poverty, focusing on clean water, agriculture, rural connectivity, and digital education.

Health and Longevity

i. Build a joint U.S.–China vaccine and treatment initiative focusing on pandemic prevention, tropical illnesses, and age-related disease.

ii. Establish a global network of biomedical research centers focusing on aging, neurodegenerative disease, and cancer.

United Nations Governance Program

Invite experts in international governance to attend a two-week-long discussion of options for strengthening or replacing the U.N. Divide the meeting into sub-groups dealing with the following topics:

i. Amend Article 2 of the U.N. charter. Make it clear that that member-state sovereignty is conditional on the protection of basic human rights. Violations of those rights will allow the U.N. to intervene to restore those rights.

ii. Abolish the veto privilege in the Security Council.

iii. Expand the rotating membership of the Council to include India, Japan, Brazil and Indonesia.

iv. Create a U.N. Panel on self-determination to investigate and rule on secession and merge requests.

v. Given the impact that local environment policies can have on global water and air patterns, set limits on what member states can do regarding the environment (e.g., tree cutting in the Amazon rainforest).

vi. Give the WHO authority to mandate and enforce global responses to pandemics.

At the end of the two weeks, subgroups will come together to report their recommendations.

4

Meditation: A Secret Weapon

In 2003 psychologist William Braud shocked the world with a book confirming his hunch that one person could influence another person physically using nothing more than his mind, i.e. without writing, speaking, visual communication, or mechanical means of any kind. The book is titled Distant Mental Influence and reports the results of 13 laboratory experiments in which the influencer (person A) is placed in one room and the person to be influenced (person B) in another. The two have never met and do not interact at any time during the experiment. Each of the 13 sessions starts with a 15-minute period in which person A tries to alter (either raise or lower) a skin-related measure of anxiety in person B; that is followed by a 15-minute period in which person A does nothing at all. That sequence is repeated twice more after which statistical comparisons are made between person B's skin-response scores during the periods when person A was trying to influence person B versus control periods when person A was doing nothing to influence person B.

The results indicate that in 10 of the 13 sessions (each using different persons A and B), person A was successful to a statistically significant degree in altering person B's skin response in the predicted direction—using nothing but his mind. Taken as a whole, the experiment offered convincing support for the claim that it is possible to influence the physical state of another person using one's mind only.[v]

People who read the book or hear about its contents differ in the way they interpret the results. There are those on the secular side who echo the author's view that the results are telling us something important about the way nature works. Then there are those who are convinced that something "miraculous" took place in the experiments. Others believe that at least some of the changes did happen but dispute the view that they were caused by gods and goddesses or represent a violation of basic physical principles. Still others believe that the results are a product of laboratory error and never really happened. Recent research in the fields of quantum physics and parapsychology suggests that even if we consider the findings "miraculous", they leave open the possibility that it is not the gods but we human beings who possess the capacity to make them happen. In The Conscious Universe scientist Dean Radin makes a case for this view.

> At a minimum, genuine psi [psychic phenomena] suggests that what science presently knows about the nature of the universe is seriously incomplete, that the capabilities and limitations of human potential have been vastly underestimated, that beliefs about the strict separation of objective and subjective are almost certainly incorrect, and that *some "miracles" previously attributed to religious or supernatural sources may instead be caused by extraordinary capabilities of human consciousness.* italics added) [vi]

While still contested by many skeptics, research in psi is creating possibilities once considered outside the limits of modern, science-based thought. As that body of research grows, more of us are willing to entertain the argument that so-called miracles may not be miraculous after all, i.e., they neither originate in heaven nor do they represent violations of basic physical laws—suggesting that these rare and wondrous events may be performed not by unseen entities in the heavens but by humans operating within what we now know to be an "entangled" universe.

Traditionally miracles were defined in two different ways, and often as both. On the one hand they were thought to be the work of a supernatural entity, typically a god or goddess who intervenes on behalf of humans here on earth. On the other, they were considered miraculous because they violated one of the physical laws that govern the universe. But both of these are under attack, whether from philosophy or science. As the world moves towards a more secular view of reality, the existence of divine beings who every now and then take pity on us humans and perform a miracle is coming into question. At the same time, our understanding of the physical universe has broadened to the point where the traditional view of human existence as a collection of separate selves who are basically unconnected to each other no longer fits the results from scientific research. Taken together these findings are shaking our faith in what was once considered inviolable truth.

Taken collectively, these recent developments have far-reaching implications for our goal of achieving world peace. They suggest that each of us has the capacity to influence others in ways that can bring that goal closer. While Braud's book says nothing about meditation per se, it seems obvious that if an untrained influencer (person A in the experiments) can change a stranger's (person B's) skin response using nothing but his or her mind, it should be possible for advanced meditators to do even more.

This opens up the possibility of changing peoples' attitudes toward members of cultures different from their own, a problem which lies at the very heart of our present paranoia. It suggests that meditators in China and the U.S., working collectively, can generate thoughts and feelings with the power to transcend the hostility toward "outsiders" that permeates both cultures.

By itself, meditation can serve as a simple and cost-effective supplement to the projects we have already discussed. I've referred to it as a secret weapon, but the truth is that it can be publicly developed and organized by both China and the United States. To facilitate that process, joint meditation sessions led by advanced practitioners can be made available on our computers and smart phones. All that requires is a little prompting from the two governments.

5

An Historical Perspective

In the light of what we know about human history, we must ask how successful we are likely to be in achieving the goals presented here, chief among them being a sustainable peace among nations. High on that list would be granting the U.N. a monopoly on military power. Other objectives include strengthening article 2 of the U.N. charter, ending the veto, and inviting India, Japan, Brazil and Indonesia into the Security Council. Altogether, these developments constitute a major reversal of global affairs as they stood in 2025. To many readers the suggestions proposed here will be seen as pie-in-the-sky—the ravings of an old man whose once-sharp brain has deteriorated into cotton candy. To see just how absurd these ideas will appear to be, take a look at the details.

For example, imagine how the public will respond when they hear of a plan to transfer all our aircraft carriers costing over a billion dollars apiece to the U.N. But don't forget the hundreds of high-tech bombers and fighters that will be going as well. Then add all the nuclear bombs that lie hidden in underground firing chutes, rocket-launchers capable of hurling explosives thousands of miles into enemy territory, and tanks so powerful they can shake off rifle fire like so many mosquitos.

If all that is not enough, add the news that the U.N. now has the authority to question and even reverse a nation's domestic policies that violate the basic rights of individuals. To add salt to these wounds, they are then told that the United States has had its veto power withdrawn and that undeveloped countries like Brazil, India and Indonesia have been invited into the Security Council.

What politicians hell-bent on getting re-elected are going to applaud such craziness? They may secretly agree that some of these changes are essential if we are to achieve lasting peace, but they can't afford to think beyond next years' primaries.

But let's be kind and assume that over the next twenty-five years, most of proposals advanced here get accepted by the general public. What could possibly explain such a dramatic reversal in thinking? Given how passionately nations have resisted criticism from outsiders, how did we manage to get nations like China and the United States to embrace a set of rules making U.N. intervention in member states' domestic affairs permissible? Such a reversal is bound to seem miraculous—unless, of course, we look deeper.

As we speculate on what the next twenty-five years will bring, one thing becomes obvious; it will take something extremely powerful to convince countries not only to give up their armies, navies and air forces, but to permit the kind of interference in domestic affairs they have always resisted. What can possibly happen in the course of those twenty-five years to permit such an extreme change of attitudes? It has to be something capable of altering the way we define ourselves and each other, i.e. something strong enough to transcend our political and religious identities and in the process reveal a truth about ourselves that has lain buried for thousands of years. Unfortunately, the only event with that kind of power is a negative one. It is WWIII.

When such a calamity comes, as it ultimately will, major cities across the globe will be hit by nuclear bombs. Millions will die—by some estimates up to 10% of the world's population, including over 100,000,000 Chinese and 30,000,000 Americans. This is slaughter on an epic scale, a form of mass murder never before seen in the history of our species. As it unfolds, our TVs will reveal figures, caught now in a radioactive vacuum, as they stumble across the street, skin sliding from their bones, lungs choking on toxic gases while in the nearby gutter, children whisper to each other as they lay dying. Just imagine this: as young boys and girls close their eyes in surrender, a barely audible cry can be heard. A plea learned in Sunday school stirs in one girl's chest, rises to her lips and spills into sound, "God, please help me." She rolls onto her side, curling her body into a fetal ball. As darkness approaches, a final whimper escapes her lips, "God, where are you?" There is no answer. And with that she is gone.

While the children lie dying, will we hear adults in New York shouting "I am an American and proud of it," or in Beijing men and women screaming "I am Chinese and you better believe it" or from the rooftops of Tehran proclaiming "Islam is the one and only true religion?" Unlikely. When your skin is hanging from your arms and legs and each breath is a struggle to stay alive, how concerned are you going to be about your identity? How upset will you be that people from "other" countries are ridiculing America or that our borders with Mexico and Canada have been violated? Again, not very. Your personal suffering is guaranteed to drown out any concern you may have about the way you identify yourself. Despite the extreme pain involved, that suffering is your gateway to a life no longer constrained by the way you identify what you are.

As you recover from the endless bombing, you may begin to see the wisdom of the policies you once found so objectionable. And that brings us to a horrible truth. To put the matter coldly, WWIII will achieve what in previous years appeared impossible. It will force us to push aside our most sacred identities—and in the process open our eyes to a new way of viewing the "others" with whom we share this globe. And that will make all the difference in the world.

Suffering is slated to play a key role in that process because it offers the possibility of redemption. Years ago, a Tibetan living in America wrote a book in which he argued that it is impossible to go deeper into oneself without suffering.[vii] He meant that even with something as trivial as a disappointment, we experience a slight withdrawal from the belief that we are in control of our destiny. He may have used the case of disappointment, but in the broader scope of things he's really talking about suffering in general. What he's pointing to is the impact that suffering has on our view of ourselves as free-willing "doers" who have the power to make things happen even when circumstances stand in the way. We suffer when things don't go our way or when we feel overwhelmed by adversity. It is during those times, he says, that we are most likely to question our life's trajectory—a questioning that when the situation permits, can lead to a re-examining of who we are and how much control we really have over events. The greater our suffering, the greater will be the potential for discovering a deeper way of looking at oneself and life in general.

Behind this line if thinking is a simple fact: if you never suffer, you are unlikely to have any interest in discovering a deeper sense of what you are. People who rarely suffer typically move along a well-trodden path, oblivious of any other way to be in the world. They have no need to question who they are, no pain to force them to ask questions about life, no desire to find a deeper sense of being. And that is the silver lining of suffering—and the ultimate reason why WWIII, despite its horror, will usher in a new chapter of hope for the citizens of planet earth.

But we have to be patient. Consider how long it took for the world to agree on something as simple as the League of Nations (a treaty the U.S. refused to ratify)—and we wouldn't have gone that far if it hadn't been for the millions of lives lost in World War I. Three decades later, we took a step closer to the ideal I'm suggesting when we created the United Nations, a more robust but militarily toothless organization where all nations were given a voice. And what prompted that modest step—WWII of course. Many more millions had to die (50-60 million by most estimates) before we were willing to transfer even a dollop of power to a centralized body with the authority to take policy positions, set some rules, and administer sanctions.

History should serve as a warning that it will most likely take a third world war and a level of suffering beyond anything the world has ever experienced before we are willing to give the United Nations (or its successor) significant military power. Getting to that final state where the U.N. is granted a *monopoly* on armed force will require that we first endure a nuclear bloodbath guaranteed to leave the world in such a ruinous state that it will take decades to make it habitable again—assuming there are enough of us left who are willing to make the effort. And then, finally, we can come to our senses and do what we should have done centuries ago—transfer all war-making weaponry to a central authority, while limiting member states to a national guard for emergencies and a police force to maintain domestic order.

When the dust of war finally settles and the vision of a world free of war is resurrected, a variety of opportunities will emerge for bringing that goal to fruition. But nothing is guaranteed. Knowing what we do about homo sapiens, habits developed over thousands of years will not be forgotten easily. Yes, there will be those who shrug off history and argue for bigger armies and more powerful bombs. But we don't have to listen. Unlike the mastodons of the last ice age who were driven to the cliff's edge by forces beyond their control, we humans have a choice. And that should give us hope.

Endnotes

i. A list of the named wars since 1900 includes the following:

World War I (1914–1918)
Russian Civil War (1917–1923)
Spanish Civil War (1936–1939)
World War II (1939–1945)
Korean War (1950–1953)
Vietnam War (1955–1975)
Arab–Israeli Wars (1948–1982, several)
Iran–Iraq War (1980–1988)
Gulf War (1990–1991)
Yugoslav Wars (1991–2001)
Rwandan Genocide and Civil War (1990–1994)
Afghanistan War (2001–2021)
Iraq War (2003–2011
Syrian Civil War
Yemeni Civil War
Russo-Ukrainian War (2014–present; full invasion 2022)
Tigray War (Ethiopia, 2020–2022)
Israel–Hamas/Gaza Wars (recurrent; 2008, 2014, 2023–24)
Nagorno-Karabakh Wars (1990s, 2020, 2023)

ii. Ralston, P. The Book of Not Knowing, North Atlantic Books, Berkely, 2010, ch. 11&12

iii. Tolle, E., The Power of Now, New World Library, Novato, 1999, ch. 9

iv. Krishnamrti, J., Ways of the Self, Krishnamurti Foundation, p. 82

v. Some replications by other researchers in the field of parapsychology have echoed his findings. Meta-analyses conducted within this community have also found small but consistent effects. However, these results remain hotly debated. Mainstream science has yet to embrace this work, citing challenges in replication, concerns over experimental rigor, and the lack of a plausible mechanism within current physical models.

In assessing the validity of Braud's thesis, it is important to note the hundreds of other laboratory studies in related areas like telepathy (reading someone's mind), clairvoyance (seeing distant objects without your biological eyes), and psychokinesis (moving physical objects with your mind only). While the results of these studies remain mixed, the majority offer evidence that under certain circumstances we humans have the potential to influence the world with our minds alone. The reason why some researchers obtain positive results while others don't remains open to question. Further studies of this intriguing issue are needed.

vi. Radin, D., The Conscious Universe, HarperCollins, New York, 2009, p324.

vii. Trungpa, C., Cutting Through Spiritual Materialism, chapter 2, Boston: Shambhala Publications, 1973.